Why Does It Happen?
OCEANS, SEAS, LAKES AND RIVERS

BABY PROFESSOR

EDUCATION KIDS

Speedy Publishing LLC
40 E. Main St. #1156
Newark, DE 19711
www.speedypublishing.com

A body of water is any significant accumulation of water, generally on a planet's surface. The term most often refers to oceans, seas, lakes, rivers and other geographical features where water moves from one place to another are also considered bodies of water.

OCEANS
&
SEAS

An ocean is a body of saline water that composes much of a planet's hydrosphere. Around 70% of the Earth's surface is covered by oceans.

Ocean is integral
to all known life,
forms part of
the carbon cycle,
and influences
climate and
weather
patterns. The
origin of Earth's
oceans remains
unknown.

Sea is often used interchangeably with "ocean" but, strictly speaking, a sea is a body of saline water partly or fully enclosed by land.

The sea moderates the Earth's climate and has important roles in the water cycle, carbon cycle, and nitrogen cycle.

LAKES

A lake is an inland body of relatively motionless water that usually has a river or stream feeding into or draining out of it.

All lakes are temporary over geologic time scales, as they will slowly fill in with sediments or spill out of the basin containing them.

There are many natural processes that can form lakes. The advancement and retreat of glaciers over millions of years can leave behind bowl-shaped depressions which fill.

Lakes can also form by tectonic related changes of the landscape, or by landslides that cause water blockages.

RIVERS

A river is a natural flowing watercourse, usually freshwater, flowing towards an ocean, sea, lake or another river.

Rivers are formed by a watershed. A watershed is an area of land where rainwater flows into valleys forming streams. These streams flow into other streams and eventually form a river.

Rivers carry water and nutrients to areas all around the earth. They play a very important part in the water cycle, acting as drainage channels for surface water.

CPSIA information can be obtained
at www.ICGtesting.com
Printed in the USA
BVHW011140040822
643794BV00012B/436

9 781682 128978